AF228906

HURRICANES

by Tammy Gagne

BrightPoint Press

San Diego, CA

© 2025 BrightPoint Press
an imprint of ReferencePoint Press, Inc.
Printed in the United States

For more information, contact:
BrightPoint Press
PO Box 27779
San Diego, CA 92198
www.BrightPointPress.com

LIBRARY OF CONGRESS CATALOGING-IN-PUBLICATION DATA

Names: Gagne, Tammy, author.
Title: Hurricanes / by Tammy Gagne.
Description: San Diego, CA: BrightPoint Press, [2025] | Series: Understanding natural disasters | Includes bibliographical references and index. | Audience: Grades 7-9
Identifiers: LCCN 2024014880 (print) | LCCN 2024014881 (eBook) | ISBN 9781678208707 (hardcover) | ISBN 9781678208714 (eBook)
Subjects: LCSH: Hurricanes--Juvenile literature.
Classification: LCC QC944.2.G34 2025 (print) | LCC QC944.2 (eBook) | DDC 363.34/922--dc23/eng20240416
LC record available at https://lccn.loc.gov/2024014880
LC eBook record available at https://lccn.loc.gov/2024014881

CONTENTS

- Hurricanes are powerful rainstorms. They form over ocean water and can travel over coastal land.

- The strength of hurricanes is categorized by wind speed. The strongest hurricanes are Category 5 storms with winds that blow 157 miles per hour (252 kmh) or more.

- Hurricane winds can be dangerous. Winds from the strongest storms can cause trees to fall and buildings to collapse.

- Floodwater from a hurricane is often the most dangerous part of the storm. People can be swept away and drown in the water.

- Hurricanes have direct effects, such as damage to buildings. They can also have indirect effects, such as a loss of jobs due to damage to businesses.

- People can prepare for a hurricane by staying informed about the approach of the storm and finding safe shelter before it hits.

- Driving through floodwater during or after a hurricane can be dangerous. A car can be swept away by as little as 12 inches (30 cm) of water.

- Rebuilding homes and lives following a hurricane can take a long time. People who feel distressed after surviving a storm may benefit from seeking mental health care.

SURVIVING HURRICANE IAN

On September 28, 2022, Hurricane Ian hit Lee County, Florida. Winds from the storm blew faster than 130 miles per hour (210 kmh). But it was the rainwater that threatened the lives of Cecilia Donald and her husband. It flooded their Fort Myers home and kept rising. They didn't know what to do. It seemed less safe to leave than it did to stay inside. Outdoors, the water could have easily swept them away.

Flash floods, which can occur after hurricanes, can take just minutes to develop into a disaster.

But being indoors was getting scarier by the minute.

Donald thought of tying herself to her husband so that they wouldn't become separated. But she couldn't find any rope. The couple climbed onto their kitchen counter. They waited for the water to stop rising. But they were afraid. Donald told her husband, "I think we're going to drown." She had been texting her daughter. The most recent update left her daughter scared. Donald's daughter said, "The last message I got from her was that the water was waist high and she didn't know if they were going to make it."[1]

Fortunately, Donald and her husband survived. But their home was badly damaged. Other people lost much more.

Phones are useful for spreading the word about a hurricane. But sometimes a hurricane can knock out cell service in an area.

Hurricane Ian killed 149 people.

Seventy-two of them lived in Lee County.

This made Hurricane Ian one of the

deadliest hurricanes in Florida history.

Palm trees grow in coastal areas where hurricanes can strike. The trees are especially flexible, which helps them survive the high winds of storms.

DEADLY DISASTERS

Hurricanes are a type of natural disaster. These strong storms form over the ocean. They have high winds and heavy rain. Sometimes hurricanes move over land. They often leave flooding behind them.

About 10,000 people are killed each year by hurricanes and tropical storms. People in a hurricane's path should find shelter. Sometimes this means staying inside until the storm has passed. But those in the path of the strongest hurricanes may need to **evacuate**.

CHAPTER ONE

WHAT ARE HURRICANES?

Powerful ocean storms are known by several names. Typhoons are the same type of weather event as hurricanes. The only difference between a typhoon and hurricane is where the storms take place. Typhoons occur in the northwest Pacific. Hurricanes happen in the North Atlantic, central North Pacific, and eastern North Pacific. Another name for these storms is cyclones.

Cyclones form over tropical and subtropical waters because those regions of the ocean are warmest.

Hurricanes begin as tropical disturbances. These systems include swirling clouds and thunderstorms. They form over tropical or **subtropical** waters. They become tropical depressions as they grow stronger. The word *depression* refers to an area of low pressure in the air. This condition helps create these storms. A tropical depression becomes a tropical storm when winds reach 39 miles per hour (63 kmh). At wind speeds of 74 miles per hour (119 kmh), the storm becomes a hurricane or typhoon.

A hurricane has three parts. The first is called the eye. This is the center of the storm. It is the calmest part. It is usually about 20 to 40 miles (30–60 km) across. Around the eye is the eye wall. This is

THE PARTS OF A HURRICANE

This diagram shows the parts of a hurricane. Scientists do not fully understand why the eyes of hurricanes form.

where the hurricane's wind and rain are
the strongest. Rain bands spin around the
wall. These bands of thunderstorms vary in
length by the size of the storm. The biggest
hurricanes have the largest bands. Some
grow up to 300 miles (480 km) long.

There are a few ways to rate a hurricane's
strength. One is called the Saffir-Simpson
Hurricane Wind Scale. It rates hurricanes
by wind speed. A hurricane with winds of
74 to 95 miles per hour (119–153 kmh) is a
Category 1 storm. A Category 1 hurricane
will likely cause some damage. Large
branches may fall from trees. The roofs of
homes may be damaged.

Winds of 96 to 110 miles per hour
(154–177 kmh) make a Category 2 storm.
This kind of storm is more likely to cause

extensive damage. These storms can bring down large trees.

Category 3 storms often cause severe damage. Entire homes can be ruined. And the risk to human life is even greater.

Beach houses are at especially high risk of damage during a hurricane.

These storms have winds that blow
111 to 129 miles per hour (178–208 kmh).
Weather experts describe the typical
damage caused by Category 4 hurricanes

On August 25, 2017, Hurricane Harvey struck Port Aransas, Texas. It was a Category 4 hurricane.

as **catastrophic**. These storms have winds of 130 to 156 miles per hour (209–251 kmh).

The strongest hurricanes are Category 5. These hurricanes have winds of 157 miles per hour (252 kmh) or more. They also cause catastrophic damage. The National Hurricane Center explains, "Most of the area [struck by a Category 5 hurricane] will be uninhabitable for weeks or months."[2]

HOW DO HURRICANES FORM?

Hurricanes are more likely to form during certain times of the year. Most hurricanes form between mid-August and mid-October. The official hurricane season in the North Atlantic lasts from June 1 to November 30. But hurricanes can happen outside this range. In the Northwest Pacific,

there is no specific cyclone season.
Storms form there at all times of the year.
In the North Indian Ocean, there are two
periods each year when cyclones are most
likely to form. One is in May. The other is
in November.

Hurricanes only form under certain
conditions. First, there must be warm ocean
water. Second, there must be humid air
above the water. When the air pressure is
just right, the humid air flows upward. A
storm forms. The warm water and moist
air feed the storm. It begins to swirl. Once
wind speeds exceed 74 miles per hour
(119 kmh), it becomes a hurricane. The
hurricane becomes stronger as it stays over
warm water. If it moves over cooler water,
the storm weakens. Hurricanes also weaken

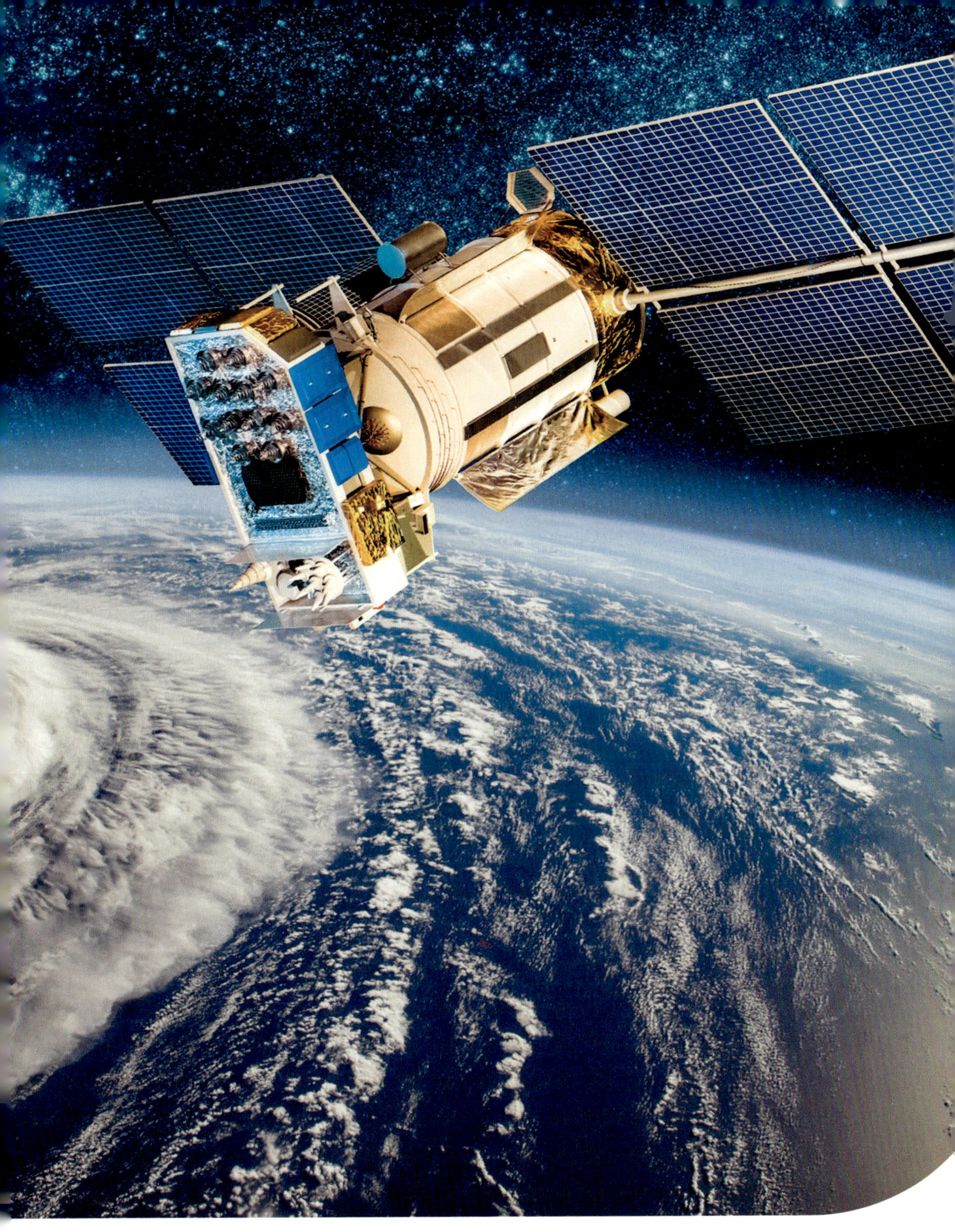

Meteorologists use satellites to track the development
and movement of hurricanes.

after they make landfall. This is because they lose the energy they drew from the warm water.

Meteorologists study hurricanes. These weather experts watch as storms develop. They try to predict how powerful a storm could get. The size of a hurricane is less telling than its wind speeds. A small storm

How Hurricanes Get Their Names

The World Meteorological Organization (WMO) names hurricanes. It has alphabetical lists of names. The lists are reused every six years. The first three storms of the 2023 hurricane season were Arlene, Bret, and Cindy. Storms are named when they reach tropical-storm level. Names of the deadliest hurricanes, such as Katrina and Floyd, are retired.

may move fast and increase in strength. A large storm may move slowly and remain weak.

Meteorologists also try to predict a hurricane's path. It can be hard to tell exactly where a hurricane will go. Global winds help move storms. Air pressure also plays a role. Experts often develop a cone-shaped range of where a storm is most likely to travel. This allows them to give warnings to people in the storm's path.

Forecasts are not perfect. Sometimes a hurricane moves in an unexpected direction. Jennifer Lowry was on vacation in Mexico in September 2014. Hurricane Odile was picking up strength nearby. She didn't expect the storm to come to her. "It was originally forecast to head West and weaken as it went out to sea," Lowry said. Then it suddenly turned in her direction. She and other guests sheltered in their hotel. Lowry explained, "Because of the initial forecast, people were not encouraged to evacuate."[3] Fortunately, none of the guests were badly hurt.

THE EFFECTS OF CLIMATE CHANGE

Climate change is the slow change of Earth's weather over time. The climate is always naturally changing. But human activity can change the climate as well. Burning fuels such as coal releases gases. Some of these gases trap the sun's heat in the **atmosphere**. These gases are known

Burning coal releases carbon dioxide, a greenhouse gas.

as greenhouse gases. Climate change has slowly raised air temperatures on Earth over the past century. This is known as global warming.

Global warming has caused ocean temperatures to rise. Warmer water has made hurricanes stronger. It has also caused ice in colder regions of the world to melt. This has raised sea levels. Rising oceans can make flooding after hurricanes more dangerous. Experts predict that hurricanes will bring more rainfall and flooding in the future.

WHO FACES THE HIGHEST RISKS?

People who live on islands or near the coast face the risk of hurricanes. Heavy rain from hurricanes leads to storm surge. This is the water from a storm that is not absorbed by soil. A storm surge can last hours or even days after a hurricane has passed. Of all US states, Florida, Louisiana, and New York have the most homes in areas prone

Storm surge can affect areas that are several miles inland.

Studies have found that people with physical disabilities are two to four times more likely to die or be injured during disasters.

to storm surge. Texas has the most homes at risk of damage from hurricane winds.

Parts of the population also face greater risk from hurricanes. Jennifer Weeks is a journalist who writes about the environment. She explains, "Even in well-prepared communities, some residents have far more resources than others to weather storms and rebuild afterward."[4] Certain people may have a harder time getting out of a hurricane's path. This raises the chance that they will be harmed in a storm. These people include older folks and unhoused people. They also include people who have disabilities or health conditions. People with lower incomes often lack insurance. They may be left unhoused if their homes are damaged in a storm.

CHAPTER TWO

THE EFFECTS OF HURRICANES

Hurricanes are dangerous in two main ways. First, these storms bring high winds. The wind can bring down trees on people. It can also cause car crashes. Wind can cause homes or other buildings to collapse. People inside a falling building can be injured or killed. The chance of a collapse rises when wind speeds reach more than 100 miles per hour (160 kmh). Certain building materials are more likely

Most trees break when wind speeds reach 94 miles per hour (150 kmh).

to fail in strong hurricane winds. Wood, for example, is much weaker than steel.

The second way a hurricane threatens people is storm surge. According to the Weather Channel, 88 percent of deaths reported from hurricanes come from rising floodwaters. The deadliest natural disaster

After the 1900 Galveston Hurricane, people built a seawall to protect the city from future hurricanes.

in US history was the 1900 Galveston
Hurricane. Galveston is a city in Texas.
The exact death toll from this storm is not
known. But as many as 12,000 people are
thought to have died. Most were probably
victims of the storm's 15-foot (4.5-m) surge.

Modern technology could have saved
many of those lives. This technology
includes satellites, television, and cell
phones. These devices make it easier to
tell people when a big storm is coming.
But lives are still lost due to storm surge
in modern times. Tropical Storm Allison hit
the Houston, Texas, area in 2001. As many
as 37 inches (94 cm) of rain fell during the
event. Forty-one people died. Twenty-seven
of the deaths were caused by flooding.

HOW HURRICANES AFFECT COMMUNITIES

Hurricanes affect communities directly and indirectly. Direct effects include loss of life, injuries, and property damage. Hurricane Andrew struck Florida in 1992. This Category 5 storm caused fifteen direct deaths in the state. It also destroyed more than 25,000 homes. Southern Florida was hit the hardest. The storm left more than 250,000 people in Dade County unhoused. Other direct effects of the storm included damage to businesses and public **infrastructure**.

The indirect effects of a hurricane are those not related to the storm itself. But they do happen in part because of it. Indirect deaths can include people who died from

The strongest hurricanes can damage infrastructure such as bridges and roads.

other injuries or illnesses. They become linked to the hurricane if it kept people from reaching a doctor. Another indirect effect could be poor farm crops due to storm damage. This effect could also lead to even more indirect effects. For example, a poor orange crop can cause people to lose their jobs picking the fruit.

Mental health problems are another effect of hurricanes. These issues may develop after a bad storm. For example, a hurricane could cause post-**traumatic** stress disorder (PTSD). PTSD can develop after a traumatic experience. It could be a

People who develop mental health disorders in the wake of a hurricane may benefit from the help of mental health professionals.

direct effect for a person who survives
a storm. It could also be an indirect effect
for those who know someone hurt by
a hurricane. Psychologist Elena Touroni
explains, "Any traumatic event—such as a
hurricane—is likely to have an impact on
a person's mental health, depending
on how safe or unsafe they felt during
the experience."[5]

The direct effects of a hurricane can
be easier to see than the indirect effects.
Researchers often study the people living
in areas hit by bad storms. Sometimes they
talk to people who were hurt or lost their
homes. This helps the researchers learn
how storms affect people. Researchers
usually begin talking to people shortly after
the storm. These experts may also follow

up months or even years later. However, people most affected by the storm often move away. This can make it hard to get a clear picture of all the long-term effects.

HURRICANES AND THE ENVIRONMENT

Hurricanes affect the environment in many ways. One of the biggest effects is erosion. The wind and water from these storms wear away the land and soil. The Florida Keys are a set of islands off the coast of the state. Hurricane Andrew stripped vegetation from some of these islands. Many mangrove trees were killed in the storm. These trees help protect the mainland from erosion. With fewer trees, more erosion will occur in coastal areas from future storms.

The exposed roots of mangrove trees support the trees and take in oxygen.

Hurricane Nicole hit Florida in 2022. It was a Category 1 storm. Nicole was far from the most powerful hurricane to strike the area. But just 6 weeks earlier, Hurricane Ian, a Category 4 storm, had made landfall. Ian had eroded the land. This allowed water from Nicole to reach farther inland. Eroded land is less stable. Ian had already damaged many houses in Volusia County. Jessica Fentress is the director of the county's coastal division. She said that after Nicole struck, several houses fell into the ocean.

Hurricanes also affect the environment by harming wildlife. Saltwater fish depend on the ocean's **salinity** for their survival. But storm-surge water reduces the ocean's salinity. Many fish die when this happens too quickly. The storm surge can also drown land animals.

Snakes on the Loose

Burmese pythons are not native to Florida. They have been spotted in the state since at least 1979. Hurricane Andrew might have released more of these snakes into the wild in 1992 by damaging breeding facilities. Since then, the snakes have bred. Experts now think it will be impossible to remove the species from the area.

Rainwater is far less salty than ocean water because salt does not rise with the water as it forms clouds.

Sometimes hurricanes have positive effects on the environment. The storms can bring much-needed rainfall to dry areas. Before Tropical Storm Debby in 2012, parts of the South were in a drought. The storm helped bring water to crops. Hurricanes can have other positive effects. An algae bloom is a rapid growth of algae in a body of

The unique color of red tide comes from a pigment in the algae called peridinin.

water. A type of algae bloom called red tide can kill marine animals. It can also make certain seafood unsafe for people to eat. Hurricanes can sometimes break up red tide in ocean water.

DEFENDING AGAINST HURRICANES

Hurricanes are powerful forces of nature. People cannot stop these storms from happening. But people can protect themselves. Staying indoors is the simplest step a person can take. Being on the beach or in the ocean are bad ideas during a hurricane. The high waves and winds from a storm may be interesting to see. But getting too close is dangerous. In 2009, Hurricane Bill killed two people who were swept out to

Governments and nonprofit organizations offer emergency shelter during severe hurricanes.

SHELTER

sea by large waves. The deaths occurred in Florida and Maine.

There are also safety steps people can take if storm surge occurs. They should move away from rising water. If people cannot evacuate, experts recommend going to the highest level of a building. But people

Contact with storm surge water should be avoided, as the water may be tainted with unsafe chemicals.

should never go into an attic that has only one exit. They could become trapped if the water rises even higher.

People should also be careful driving during and after a hurricane. High winds and heavy rain can make it hard to see ahead. Storms can bring down trees and telephone poles. These objects can hit vehicles or block roads. Rising water can also be a big problem. Experts warn drivers to turn around if they come to a flooded part of the road. Just 12 inches (30 cm) of moving water is enough to sweep away a large vehicle.

Julie Munger trains people to rescue others from water emergencies. She says that the best course of action is never driving through standing water of

Deep water can stall a car's engine, leaving it stuck in the flooded portion of a road.

any depth. She says, "Everybody tends to underestimate the force of the water."[6]

STORM PREP

People in a hurricane's path can take steps to prepare. They should consider stocking up on supplies. These include food, bottled water, flashlights, and batteries. Stores often

close during the worst parts of a hurricane.
And some storms can leave homes without
electricity for hours or days.

Another way to prepare for a hurricane
is by listening to news reports. A radio
powered by batteries can come in handy
if the power goes out. Authorities may
issue hurricane watches and warnings.
A watch means that hurricane conditions
are possible. A hurricane warning is more
serious. It means that weather experts
expect the storm to reach the area.

People may need to make further
preparations for severe hurricanes. They
might cover windows or glass doors with
plywood. They might also remove objects
that strong winds could move. Some people
close to the coast even stack sandbags

in front of their homes. This can help keep floodwaters from entering the building.

Knowing when to seek shelter is important. News reports will tell viewers ahead of a hurricane where to go if they need shelter. Schools, churches, and other large buildings not in danger of flooding may be used as shelters. Sometimes people can choose for themselves whether to leave. But occasionally the government

Evacuation Routes

In areas prone to hurricanes and other natural disasters, officials may mark evacuation routes. Officials may post road signs on the routes. These signs are often blue and white with the words *evacuation route* clearly displayed. Following a marked route can make the process easier and safer for drivers.

People stack sandbags to protect buildings and other areas from rising floodwaters.

issues a **mandatory** evacuation. This means everyone within the area should leave as soon as possible.

Hurricane Katrina struck the southeastern United States in August 2005. It killed 1,392 people. This is the second-highest death toll from a US natural disaster in modern history. According to the Federal Emergency Management Agency, the number might have been much lower if the evacuation had been managed better. Many people blamed the government. They said it did not respond as well as it could have. New Orleans, Louisiana, suffered greatly during the hurricane. Afterward, the state spent lots of money on rebuilding. New Orleans

New Orleans is surrounded by a system of levees, which repel storm surge after a hurricane. However, they can fail, as they did after Hurricane Katrina struck.

updated its evacuation plans. It said it would evacuate people from threatened areas. This would include making arrangements for people who need extra help.

The deadliest natural disaster in modern US history was Hurricane Maria. This Category 5 storm struck Puerto Rico in 2017. It also affected nearby islands. It killed 2,975 people. Many people said that the government responded too slowly to the storm. Quick relief is necessary for areas recovering from hurricanes.

DEALING WITH THE AFTERMATH OF HURRICANES

Hurricanes are some of the most destructive of all natural disasters. It takes a lot of time, money, and effort to recover from these storms. Hurricane survivors may need to completely rebuild their homes. They may also need to recover from the trauma of a hurricane.

Bobby and Jodi Mann live in Lee County, Florida. The Manns' home in St. James City was badly damaged by Hurricane Ian in 2022. Bobby described the damage. He said, "It was over 4 feet [1 m] of ocean throughout our house, churning like a washing machine."[7] The couple had trouble finding a contractor to rebuild their house after the storm. So many homes needed repairs that waitlists were months long. The Manns decided to rebuild themselves.

In the meantime, the Manns moved into a trailer home. They parked it in their driveway while they worked on their house.

Sometimes rebuilding a home after a natural disaster is less expensive than moving to a new home.

In August 2023, the house still had no roof. Instead, a tarp stretched across the top of the house. As they rebuilt, they worried that another storm would strike their town.

Dr. Ronald Smallwood is a psychiatrist in Florida. He says that people need to take care of their mental health after events such as hurricanes. It is common for people to feel difficult emotions after

**Surviving a hurricane can be a traumatic experience,
but there is help available for people who have done so.**

natural disasters. These feelings can include hopelessness, exhaustion, and despair. He advises, "If the suffering continues to get worse and doesn't resolve with time, then it may be a good idea to try and obtain professional help."[8]

Hurricanes cannot be prevented. But people who live in areas most prone to these storms can prepare for them. Knowing how hurricanes work helps people in their paths stay safe as the storm passes.

GLOSSARY

atmosphere
the layer of gases that surround Earth

catastrophic
causing devastating damage

evacuate
to leave an area to avoid danger

infrastructure
structures such as bridges and highways used by large numbers of people in an area

mandatory
required by law

meteorologists
scientists who study weather

salinity
the salt content of water

subtropical
the area between tropical and temperate climate zones

traumatic
extremely disturbing

SOURCE NOTES

INTRODUCTION: SURVIVING HURRICANE IAN

1. Quoted in Norah O'Donnell and Angel Canales, "Survivors of Hurricane Ian Share Harrowing Stories: 'I Think We're Going to Drown,'" *CBS News*, September 29, 2022. www.cbsnews.com.

CHAPTER ONE: WHAT ARE HURRICANES?

2. "Saffir-Simpson Hurricane Wind Scale," *National Hurricane Center and Central Pacific Hurricane Center*, n.d. www.nhc.noaa.gov.

3. Quoted in "Stories from Hurricane Survivors," *National Weather Service*, n.d. www.weather.gov.

4. Jennifer Weeks, "Understanding Hurricane Risks: 5 Essential Reads," *The Conversation*, June 1, 2018. www.theconversation.com.

CHAPTER TWO: THE EFFECTS OF HURRICANES

5. Quoted in Adam England, "Repeated Exposure to Hurricanes Can Be Detrimental to Mental Health," *Verywell Mind*, July 30, 2022. www.verywellmind.com.

CHAPTER THREE: DEFENDING AGAINST HURRICANES

6. Quoted in Susan Shain, "What to Do When There's a Flash Flood Warning," *New York Times*, February 2, 2024. www.nytimes.com.

7. Quoted in Christina Zdanowicz and Kara Nelson, "They Lost Everything in Hurricane Ian and Are Bracing for Idalia. Here's What They Want You to Know," *CNN*, August 29, 2023. www.cnn.com.

8. Quoted in "After the Storm: How to Deal with Mental Health Challenges," *Lee Health*, October 11, 2022. www.leehealth.org.

FOR FURTHER RESEARCH

BOOKS

Cody Crane, *All About Hurricanes*. New York: Children's Press, 2021.

Sue Bradford Edwards, *Tornadoes*. San Diego, CA: BrightPoint Press, 2025.

Lauren Tarshis, *I Survived the Galveston Hurricane, 1900*. New York: Scholastic, 2021.

INTERNET SOURCES

"Hurricanes," *National Geographic Kids*, n.d. https://kids.nationalgeographic.com.

"Hurricanes," *National Oceanic and Atmospheric Administration*, May 1, 2020. www.noaa.gov.

"Hurricanes," *Ready*, March 21, 2024. www.ready.gov.

WEBSITES

American Red Cross
www.redcross.org

The American Red Cross provides relief after disasters, including hurricanes.

National Hurricane Center
www.nhc.noaa.gov

The National Hurricane Center shares hurricane forecasts and data.

National Weather Service
www.weather.gov

The National Weather Service gives information about past and present weather events in the United States and its territories.

INDEX

IMAGE CREDITS

Cover: © FotoKina/Shutterstock Images

5: © 3dmotus/Shutterstock Images

7: © Brisbane/Shutterstock Images

9: © Kostenko Maxim/Shutterstock Images

10: © Photobank.kiev.ua/Shutterstock Images

13: © SergeyIT/Shutterstock Images

15: © Evgeniyqw/Shutterstock Images

17: © Leonard Zhukovsky/Shutterstock Images

18: © Roschetzky Photography/Shutterstock Images

21: © Andrei Armiagov/Shutterstock Images

25: © Vikojam/Shutterstock Images

27: © Ccpixx Photography/Shutterstock Images

28: © Andrey_Popov/Shutterstock Images

31: © egd/Shutterstock Images

32: © John Blottman/Shutterstock Images

35: © Samuel Acosta/Shutterstock Images

36: © fizkes/Shutterstock Images

39: © Chad Zuber/Shutterstock Images

41: © lee.f_oto/Shutterstock Images

42: © Alfred Rowan/Shutterstock Images

45: © Jewelzz/Shutterstock Images

46: © IrinaK/Shutterstock Images

48: © Igor Corovic/Shutterstock Images

51: © Jane H. York/Shutterstock Images

52: © Chuck Wagner/Shutterstock Images

55: © Artush/Shutterstock Images

56: © Marjan Apostolovic/Shutterstock Images

ABOUT THE AUTHOR

Tammy Gagne is an author and editor who specializes in nonfiction. She has written hundreds of books for both children and adults. She lives in northern New England with her husband, son, and dogs.